MW00682178

FLYING ANIMALS

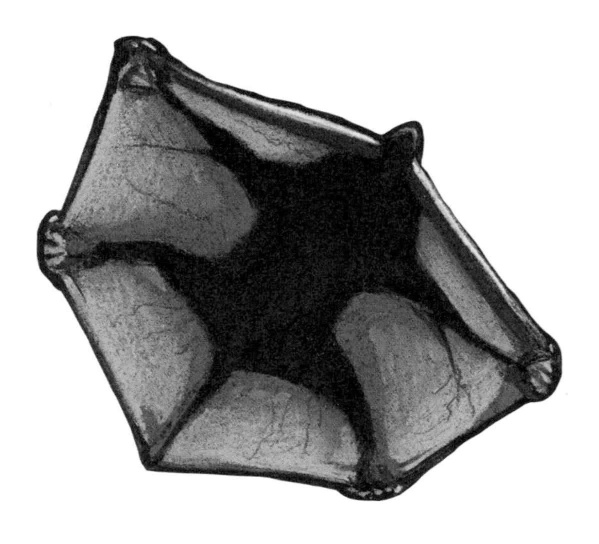

By Patricia Lantier-Sampon
Illustrated by Jeff Meyer

Gareth Stevens Publishing
MILWAUKEE

For a free color catalog describing Gareth Stevens' list of high-quality books, call 1-800-341-3569 (USA) or 1-800-461-9120 (Canada).

Library of Congress Cataloging-in-Publication Data

Lantier-Sampon, Patricia.
Flying animals / by Patricia Lantier-Sampon; illustrations by Jeff Meyer (Jeffrey D.).
p. cm. — (Wings)
Includes index.
ISBN 0-8368-0540-2
1. Animals—Juvenile literature. 2. Animal flight—Juvenile literature. [1. Animals. 2. Flight.] I. Meyer, Jeff, ill. II. Title. III. Series: Lantier-Sampon, Patricia — Wings.
QL49.L325 1994
591—dc20 91-50346

Edited, designed, and produced by
Gareth Stevens Publishing
1555 North RiverCenter Drive, Suite 201
Milwaukee, Wisconsin 53212, USA

Designer: Kristi Ludwig

Printed in the United States of America

1 2 3 4 5 6 7 8 9 99 98 97 96 95 94

Contents

People have always watched animals fly. Their movements inspire us to conquer the sky.

Pterosaurs

Fur-covered pterosaurs had wings made for gliding. But now they're extinct — unless they're all hiding!

Bats

Bats live in roosts and promote pollination. They find their direction through echolocation.

Flying Squirrels

Small flying squirrels make long, skimming glides.

They have special membranes attached to their sides.

Flying Fish

The grand flying fish has oversized fins. In races with friends, it undoubtedly wins!

Flying Snakes

The strange flying snake lives mainly in trees. It eats bats and lizards and glides with great ease.

Flying Lizards

Bizarre flying lizards have a dewlap and wattle.

They're just not the kind of pets
you can coddle!

Colugos

The little colugo is a nocturnal creature. Its thirty-four teeth are a curious feature.

Flying Frogs

Frisky flying frogs have enormous webbed feet. Their parachute movements are really quite neat.

Flying Phalangers

Flying phalangers sport feather-like tails. And they have special flaps that serve as wind sails.

Flying People

Skydivers, swing artists, hang gliders, too. All try to soar through the sky like birds do!

People have always wished they could fly. Wouldn't it be fun to have wings?

Glossary

dewlap: loose skin that hangs under the neck of an animal.

echolocation: a way to find, or locate, things by using sound waves.

membrane: a thin layer of skin.

nocturnal: active during the night.

roost: a resting place or perch.

wattle: a thick fold of skin around the head, throat, or neck of an animal.

Index